Felix Ale

The Noble Role of Space Journalism in Global History

GRIN Publishing

Bibliographic information published by the German National Library:

The German National Library lists this publication in the National Bibliography; detailed bibliographic data are available on the Internet at http://dnb.dnb.de .

Imprint:

Copyright © 2015 GRIN Verlag GmbH
Print and binding: Books on Demand GmbH, Norderstedt Germany
ISBN: 978-3-656-90941-5

This book at GRIN:

http://www.grin.com/en/e-book/293438/the-noble-role-of-space-journalism-in-global-history

GRIN - Your knowledge has value

Since its foundation in 1998, GRIN has specialized in publishing academic texts by students, college teachers and other academics as e-book and printed book. The website www.grin.com is an ideal platform for presenting term papers, final papers, scientific essays, dissertations and specialist books.

Visit us on the internet:

http://www.grin.com/

http://www.facebook.com/grincom

http://www.twitter.com/grin_com

THE NOBLE ROLE OF SPACE JOURNALISM IN GLOBAL HISTORY

BEING PAPER SUBMITED
BY
FELIX ALE
ATLANTIC INTERNATIONAL UNIVERSITY

The media is a powerful tool for distribution of information. As the mouth piece of society, the fourth estate is responsible for ensuring that the audience is kept abreast of happenings in the society. This has enabled the media to possess the power of agenda setting across the globe, dictating the nature of developments in the society. Usually, the media dictates the development of ideas and innovations by highlighting the benefits of such developments. To this end, the media can be applauded for its beneficence to the society, in not only conveying messages to geographically dispersed audience but also prompting action and intervention through opinion formation and reinforcement of issues that need these. Space technology is one aspect of human development that has benefited immensely from media focus in Science and technology. Having emerged just in the 1950s, media coverage and analysis led to many countries considering implementing this technology. While the development of space science might have been encouraged by global hegemonic and war strategies, the media played a bigger role in creating awareness about the technology.

From time immemorial, journalism has been cited as one of the sources of inspiration in the societies. The inspirations come from the fact that journalists help drive specific agendas in the society. Technology-wise, the inspirations of the media can be seen from the manner in which they have taken advantage of wars and made reports and analysis that prompted the adoption of certain war technologies into civil use. After the First World War, the advent of industrialization ushered countries to a new approach of winning economic competence. The civilized European powers and United States of America engaged on economic propulsion and scientific discoveries, which led to development of mass destruction weapons and fast warships as each country funded weaponry manufacturers and scientists. While there are many technological development that have emerged from war and which have been adopted for civil use, space technology stand out as one of the post-second war technologies that has been shaped through journalism.

At the height of the cold war in 1957, the Soviet Union launched the first ever satellite. The satellite was named Sputnik. Sputnik was designed to orbit the earth and send sensitive intelligence information to the Soviets. The Sputnik was small in size, almost the size of a beach ball, and was not able to send meaningful information back to the earth. However, the launch of the Sputnik had a profound impact on the priorities of citizens and governments around the world (McQuaid, 2007). The Soviet engineers who developed the Sputnik wanted to ensure that the satellite could be seen and heard by people from all parts of the world. While Sputnik was intended to improve communication by the soviets, it has a ripple effect that saw many nations developing and launching their own satellites.

The space age was ushered in by the cold war era. However, the media spurred it through reporting and analysis about the tussled between the United States and the Soviet Union. The Sputnik made space exploration possible and prompted the United States to embrace the technology. During the cold war, the soviets and the United States engaging in a way that was less physical but was more about threats and threat-making (Komska, 2011). Given that the soviets had been the first to come up with the idea of space crafts, the united states was obligated to follow , and probably make it better. Thus with the launch of the Sputnik, the United States and the Soviet Union entered into a new race of space technology. The reports by the media on the space capabilities of each side led to the drive for nations to develop their space capacities. Hence the crucial beginning of Space Journalism in Media practice.

One of the ways through which journalists were involved in the space development was through reporting the space technological developments in various countries. During the initial stages of space science, the focus was mainly on the United States and the Soviet Union. With the two countries competing for global supremacy, the media in each country ensured that they reported any new developments or intelligence leaks about impending space technological developments by the "enemy" country (PBS, 2009). For example, the reaction of the United States towards Sputnik was to develop their own space technology to counter the Soviets. The Sputnik was considered an 'enemy satellite' (Shreve, 2003). This meant the United States had to do something to counter the threat of the enemy. To counter the soviets, the United States initiated several space programs that would develop much stronger satellites. The concern at the time was that the soviets had their satellites over the airspace of the United States and there was nothing that the United States could do about it. The United States believed that the soviets were able to get vital information about the United States that would enable them to launch nuclear bombs against the United States. However, what the United States failed to realize at the time was that since the soviets had rockets powerful enough to send a spacecraft into the space, they were able to launch a nuclear bomb against the United States at will.

The media hype about the space technologies and the capabilities of the satellites coupled with the cold war objectives led the United States to create the National Aeronautics and Space Administration (NASA). However, the initial attempts by the United States to launch satellites into the space ended up in spectacular explosions with journalists taking an active role in the discussions before and after the attempts. Soon, however, the government realized the need for building space capsules, rockets and satellites (Frederick, 2011). Besides, the need to hire astronauts to become spacemen emerged.

Ultimately, the space development turned into some kind of contest with journalists reporting on every new development. For American journalists, the driving factor was that the Soviets were wining the space battle. Advancements in technology and space explorations that we see today can be attributed to the attempts by the United States to catch up with the Soviet Union during the space age. The space race altered the educational system and the imagination of individuals in Nebraska and other regions of the United States. Shortly after the soviets had lunched the Sputnik, American journalists begun calling on the American education system to put more emphasis on science and math. This led to the government of United States pumping more money to enable schools and educators focus on math and science (Geppert, 2012). This means most students enrolled in math and science courses.

The greatest highlight of the space age is arguably the Apollo programs by the United States. With the Soviets having become the first nation to send a satellite into space, the United States was determined to become the first nation to send a human being to the space. Thus the most famous outcome of the Apollo programs was the *Apollo 11*. The *Apollo 11* was the spacecraft that carried the first human beings to the moon, astronauts Michael Collins and Buzz Aldrin, in 1969 (PBS, 2009). This was followed by a number of other space explorations to the moon, most of which were under the same Apollo program. The main aim of the American space explorations during the cold war was to study the planets and their satellites. Eventually, space explorations became influential in providing information about the solar system, and the earth in particular (The cold war Museum, 2010). Over the recent times, the most historic space event is the establishments of the Hubble Space Telescope which gave scientists the ability to get more insights into the solar system.

The greatest contribution of journalists in the development of the space science was the focus on the technology. While the United States and the Soviet Union were focusing on the need to become the "first to", the media was reporting on how the space technologies were developing. Thus besides the journalists reporting the launching of satellites to the space, the most important highlights of the space age were the developments of rockets. Thus the media was instrumental in defining the space technology. The space age was definitely defined by the creation and use of rockets. Indeed, most other space technologies that people use today were developed during the space age. Most spacecrafts, for example, needed ceramics that could withstand the extreme heat (Macdonald, 2008). These ceramics first emerged during the space age.

Interestingly, the media has been on the forefront of adopting space technologies such as satellites. Satellites are some of the most commonly used and advanced technologies in recent times. Satellites are space-based radio systems that have been used for various purposes. Satellites orbit the earth and function to provide 24 hour coverage of the earth. Perhaps the most effective use or contributions of satellites has been in the communication sector. Satellites have been used for a long time for several purposes. Steeliest have aided telephone communication by connecting people from remote parts of the world. Fix-point telephone system has been used to relay telephone communication to earth stations without the net of wired connection (Komska, 2011). Besides, satellites are currently used for radio and television transmission through direct broadcast satellites and fixed service satellite (Komska, 2011). Today, satellites are used to monitor weather patterns by providing pictures necessary to study state of the atmosphere. For example, hurricanes are today observed and detected before they ambush people living along the coasts. Scientist has also used satellites for observing the activities on the earth such as the extent of global warming in various parts of the world forest cover and the nature of agriculture in the world. From all these observations, it can be said that satellite technology has been beneficial to mankind and the current technological developments would not have been available without satellite technology.

From the development of space technologies, it can be said that journalists were effective in using agenda setting in encouraging both the United States and the Soviet Union to increase their spending in space technologies. The media uses agenda setting to influence the perception of the audience. Agenda setting refers to the fixing of a media message and its frequency in order to create a sense of urge or importance of the news. Frequency of a news article or coverage in the media will heighten awareness and interest which consequently lead to action, and this becomes the main purpose of the media in this sense (McQuaid, 2007). However, the media may use agenda setting inappropriately to spur the revolution. For example, despite the fact that initial space attempts by the United States had flopped, the media focus on American superiority in technology finally made it possible for the United States to come up with better space technologies.

Following the end of the cold war, the biggest concern about nuclear weapons has not been the use of nuclear weapons against specific targets, but the fate of the nuclear stock piles in Russia, The United states, China, Israel, India, Pakistan, France and the UK. Again, the media was called into action. Having seen the destructive effects of nuclear weapons, journalists had taken the forefront in advocating for safer storage of nuclear arsenals. Of most concern have been the loose nukes in Russia and other former Soviet Union states where there were thousands assembled nuclear weapons. The concern is that poorly protected nuclear weapons could find their way into the hands of terrorists and criminals. However, there has been substantial reduction in the number and usage of nuclear weapons. This has resulted from various reasons including individual decisions by specific countries to reduce the number of their nuclear arsenal, international treaties and barriers brought about by international business and corporations.

In conclusion, it is indeed an undisputable fact that journalism played an important role in the development of the space technology in global affairs. Through reporting on the various cold war and other related events, journalists initiated a space age that was marked by intense space technology competition between the United States and the Soviet Union. The launching of the Sputnik marked the beginning of the space race that saw rapid technological developments in the various parts of the world. The United States felt threatened by the Sputnik and made the necessary steps to counter the "threat". But while the United States was trying to neutralize the effects of the Sputnik, they realized new space frontier that could be explored. And in 1969, the United States pioneered the cause of history to become the first country in the world to send astronauts to the moon. This development caused a media frenzy of its kind in the United States and many other countries in the world. Besides, the space race led to the developments of various space-related technologies such as rocketary that have become common with armies for launching missiles. All these development were soured by media reporting and analysis in one way or another.

Therefore, space journalism remain a very strategic part of the media that has played a very crucial role in global affairs and its noble contributions in the promotion of Space Science and technology will continue to loom large in human history, particularly in the technological revolution across the globe.

References

Frederick, L. A. (2009). Deterrence and Space-Based Missile Defense. *Air & Space Power Journal, 23*(3), 107-118.

Geppert, A. T. (2012). Introduction rethinking the space age: Astroculture and technoscience. *History & Technology, 28*(3), 219-223.

Komska, Y. (2011). Ruins of the Cold War. *New German Critique,* (112), 155-180.

Macdonald, F. (2008). Space and the Atom: On the popular geopolitics of cold war rocketry. *Geopolitics, 13*(4), 611-634.

McQuaid, K. (2007). Sputnik reconsidered: Image and reality in the early space age. *Canadian Review of American Studies, 37*(3), 371-401.

PBS (2009). *Space race timeline.* Retrieved from http://www.pbs.org/wgbh/nova/astrospies/time-nf.html

Shreve, B. G. (2003). The US, the USSR, and Space Exploration, 1957-1963. *International Journal on World Peace, 20*(2), 67-83.

The cold war Museum (2010). *The space race.* Retrieved from http://www.coldwar.org/articles/60s/space_race.asp